I0475771

U.S. ENVIRONMENTAL PROTECTION AGENCY

OFFICE OF INSPECTOR GENERAL

Fiscal Years 2012 and 2011 (Restated) Financial Statements for the Pesticide Registration Fund

Report No. 14-1-0042   December 17, 2013

Report Contributors:
Paul Curtis
Robert Smith
Kelly Bonnette
Edgar Dumeng
Sabrina Jones
Carol Kwok
Mairim Lopez
Claire McWilliams
Guillermo Mejia
Demetrios Papakonstantinou
Myka Sparrow
Lynda Taylor

Abbreviations

EPA	U.S. Environmental Protection Agency
EPM	Environmental Programs and Management
FIFRA	Federal Insecticide, Fungicide, and Rodenticide Act
FMFIA	Federal Managers' Financial Integrity Act
FY	Fiscal Year
OCFO	Office of the Chief Financial Officer
OIG	Office of Inspector General
OMB	Office of Management and Budget
OPP	Office of Pesticide Programs
PRIA	Pesticide Registration Improvement Act

At a Glance

Why We Did This Review

The Pesticide Registration Improvement Act requires that we perform an annual audit of the Pesticide Registration Fund (known as the PRIA Fund) financial statements.

To expedite the registration of certain pesticides, Congress authorized the U.S. Environmental Protection Agency (EPA) to assess and collect pesticide registration fees. The fees collected are deposited into the PRIA Fund. The agency is required to prepare financial statements that present financial information about the PRIA Fund. PRIA also requires the establishment of decision time review periods for pesticide registration actions, and requires the Office of Inspector General to perform an analysis of the agency's compliance with those review periods.

This report addresses the following EPA theme:

- *Embracing EPA as a high performing organization.*

For further information, contact our public affairs office at (202) 566-2391.

The full report is at:
www.epa.gov/oig/reports/2014/20131217-14-1-0042.pdf

Fiscal Years 2012 and 2011 (Restated) Financial Statements for the Pesticide Registration Fund

EPA Receives an Unqualified

We rendered an unqualified, or clean, opinion on the EPA's Pesticide Registration Fund financial statements for fiscal years (FYs) 2012 and 2011 (restated), meaning they are fairly presented and free of material misstatement.

Internal Control Material Weaknesses Noted

We noted two material weaknesses in internal controls.

The agency corrected material misstatements due to weaknesses in internal controls we identified.

- EPA materially overstated the expenses from other appropriations that support the PRIA fund. This occurred because the agency does not have an effective system to accurately accumulate and report costs incurred by other appropriations in support of PRIA Fund activities. This overstatement resulted in a material overstatement of the total costs of the PRIA Fund by $14.1 million in FY 2012 and $1.7 million in FY 2011.

- EPA materially understated the PRIA fund payroll liabilities covered by budgetary resources as well as related payroll expense included in gross costs. The agency's practice of transferring employees and related expenses and liabilities from PRIA to the Environmental Programs and Management Fund for cash flow reasons led to the understatement. The FY 2011 payroll liabilities covered by budgetary resources for PRIA was $500,000, while the FY 2012 payroll liabilities covered by budgetary resources was zero.

Compliance with Decision Time Review Periods

The agency was in compliance with applicable laws and regulations.

Recommendations and Planned Agency Corrective Actions

The agency agreed with our findings and our recommendations. The agency corrected the financial statements to reflect the proper expenses paid by other appropriations and to reflect the proper payroll liability amounts. The agency will also develop a process to ensure accurate allocations of expenses from other appropriations that support the PRIA fund and carefully review and comment on the draft and final versions of the PRIA financial statements prior to their submission to the Office of Inspector General. The agency will also closely monitor the payroll amounts.

UNITED STATES ENVIRONMENTAL PROTECTION AGENCY
WASHINGTON, D.C. 20460

OFFICE OF
INSPECTOR GENERAL

December 17, 2013

MEMORANDUM

SUBJECT: Fiscal Years 2012 and 2011 (Restated) Financial Statements for the
Pesticide Registration Fund
Report No. 14-1-0042

FROM: Paul C. Curtis
Director, Financial Statement Audits

TO: Jim Jones, Assistant Administrator
Office of Chemical Safety and Pollution Prevention

Maryann Froehlich, Acting Chief Financial Officer

Attached is our report on the U.S. Environmental Protection Agency's (EPA's) fiscal years 2012 and 2011 (restated) financial statements for the Pesticides Registration Fund, conducted by the EPA Office of Inspector General (OIG). This audit report represents the opinion of the OIG, and the findings in this report do not necessarily represent the final EPA position. EPA managers, in accordance with established EPA audit resolution procedures, will make final determinations on the findings in this audit report. Accordingly, the findings described in this audit report are not binding upon EPA in any enforcement proceeding brought by EPA or the Department of Justice. This report will be available at http://www.epa.gov/oig.

In accordance with EPA Manual 2750, we are closing this report on issuance in our tracking system. You should track progress of your corrective actions in the Management Audit Tracking System.

If you or your staff have any questions regarding this report, please contact Richard Eyermann, Acting Assistant Inspector General for Audit, at (202)566-0565 or eyermann.richard@epa.gov; or Paul Curtis, Director, Financial Statement Audits, at (202)566-2523 or Curtis.Paul@epa.gov

Table of Contents

Inspector General's Report on the Fiscal Years 2012 and 2011 (Restated) Financial Statements for the Pesticide Registration Fund

Attachments

Appendices

Inspector General's Report on the Fiscal Years 2012 and 2011 (Restated) Financial Statements for the Pesticide Registration Fund

The Administrator
U.S. Environmental Protection Agency

We have audited the Pesticide Registration Fund (known as the PRIA Fund) balance sheet as of September 30, 2012 and 2011 (restated), and the related statements of net cost, changes in net position, and budgetary resources for the years then ended. These financial statements are the responsibility of U.S. Environmental Protection Agency (EPA) management. Our responsibility is to express an opinion on these financial statements based upon our audit.

We conducted our audit in accordance with the generally accepted auditing standards; the standards applicable to financial statements contained in *Government Auditing Standards*, issued by the Comptroller General of the United States; and Office of Management and Budget (OMB) Bulletin No. 07-04, *Audit Requirements for Federal Financial Statements, as Amended.* These standards require that we plan and perform the audit to obtain reasonable assurance as to whether the financial statements are free of material misstatements. An audit includes examining, on a test basis, evidence supporting the amounts and disclosures in the financial statements. An audit also includes assessing the accounting principles used and significant estimates made by management, as well as evaluating the overall financial statement presentation. We believe that our audit provides a reasonable basis for our opinion.

The agency has restated the PRIA Fund financial statements for fiscal year (FY) 2011 due to material errors in the computation of expenses from other appropriations that support PRIA Fund activities. There errors resulted in an overstatement of these expenses by $14.1 million in FY 2012 and $1.7 million in FY 2011. The agency has restated the FY 2011 financial statements to reflect the decrease of the expenses from other appropriations that support the PRIA Fund and made corresponding adjustments to the other related accounts. Due to material errors found in the computation of the expenses from other appropriations that support PRIA Fund activities and other related accounts, our report on the PRIA Fund FY 2011 financial statements, issued on June 6, 2012, is not to be relied upon. That report is replaced by this report on the restated FY 2011 PRIA Fund financial statements. We report the internal control deficiency that resulted in the material errors as a material weakness in the Internal Control section of this report.

In our opinion, the financial statements, including the accompanying notes, present fairly, in all material respects, the assets, liabilities, net position, changes in net position and budgetary resources of the PRIA Fund, as of and for the years ending September 30, 2012 and 2011, as restated, in conformity with accounting principles generally accepted in the United States of America.

Evaluation of Internal Controls

As defined by OMB, internal control, as it relates to the financial statements, is a process, affected by the agency's management and other personnel, that is designed to provide reasonable assurance that the following objectives are met:

Reliability of financial reporting – Transactions are properly recorded, processed and summarized to permit the preparation of the financial statements in accordance with generally accepted accounting principles, and assets are safeguarded against loss from unauthorized acquisitions, use or disposition.

Compliance with applicable laws, regulations and governmentwide policies – Transactions are executed in accordance with laws governing the use of budget authority, governmentwide policies, laws identified by OMB, and other laws and regulations that could have a direct and material effect on the financial statements.

In planning and performing our audit, we considered the EPA's internal control over the Pesticide Registration Improvement Act (PRIA) financial reporting by obtaining an understanding of the agency's internal controls, determining whether internal controls had been placed in operation, assessing control risk, and performing tests of controls. We did this as a basis for designing our auditing procedures for the purpose of expressing an opinion on the financial statements and to comply with OMB audit guidance, not to express an opinion on internal control. Accordingly, we do not express an opinion on internal control over financial reporting or on management's assertion on internal controls included in Management's Discussion and Analysis. We limited our internal control testing to those controls necessary to achieve the objectives described in OMB Bulletin No. 07-04, *Audit Requirements for Federal Financial Statements, as Amended.* We did not test all internal controls relevant to operating objectives as broadly defined by the Federal Managers' Financial Integrity Act (FMFIA) of 1982, such as those controls relevant to ensuring efficient operations.

Our consideration of the internal controls over financial reporting would not necessarily disclose all matters in the internal control over financial reporting that might be significant deficiencies. Under standards issued by the American Institute of Certified Public Accountants, a significant deficiency is a deficiency, or combination of deficiencies, in internal controls that is less severe than a material weakness, yet important enough to merit attention by those charged with governance. A material weakness is a deficiency, or combination of deficiencies, in internal controls, such that there is a reasonable possibility that a material misstatement of the entity's financial statements will not be prevented, or detected and corrected, in a timely basis. Because of inherent limitations in internal controls, misstatements, losses or noncompliance may nevertheless occur and not be detected. We noted two matters involving the internal controls and their operations that we considered to be a material weakness.

Material Weaknesses

Material weaknesses noted are summarized below and detailed in attachment 1.

EPA Materially Overstated the Expenses From Other Appropriations That Support PRIA. EPA materially overstated the expenses from other appropriations that support the PRIA fund. This occurred because the agency does not have an effective system to accurately accumulate and report costs incurred by other appropriations in support of PRIA Fund activities. This overstatement of the expenses from other appropriations resulted in a material overstatement of the total costs of the PRIA Fund by $14.1 million in FY 2012 and $1.7 million in FY 2011. This overstatement could impact the opinion on the financial statements and reliance on reported PRIA financial information.

EPA Understated PRIA Payroll Liabilities Covered by Budgetary Resources. EPA materially understated the PRIA fund payroll liabilities covered by budgetary resources as well as related payroll expense included in gross costs. The agency's practice of transferring employees and related expenses and liabilities from PRIA to the Environmental Programs and Management (EPM) Fund for cash flow reasons led to the understatement. The FY 2011 payroll liabilities covered by budgetary resources for PRIA was $500,000, while the FY 2012 payroll liabilities covered by budgetary resources was zero. Such understatements could impact the opinion on the financial statements and reliance on reported PRIA financial information.

Comparison of EPA's FMFIA Report and With Our Evaluation of Internal Controls

OMB Bulletin No. 07-04, *Audit Requirements for Federal Financial Statements, as Amended*, requires us to compare material weaknesses disclosed during the audit with those material weaknesses reported in the agency's FMFIA report that relate to the financial statements and identify material weaknesses disclosed by the audit that were not reported in the agency's FMFIA report.

For financial statement, audit and financial reporting purposes, OMB defines material weaknesses in internal control as a deficiency or combination of deficiencies in internal control, such that there is a reasonable possibility that a material misstatement of the financial statements will not be prevented or detected and corrected on a timely basis. The agency did not report any material weakness for FY 2012 impacting the PRIA Fund; however, we identified a material weakness with the agency's reporting payroll and benefit payable. Details concerning these material weaknesses are in attachment 1.

Tests of Compliance With Laws and Regulations

In accordance with PRIA, the Administrator is required to publish a schedule of decision time review periods for pesticide registration actions and corresponding registration fees in the Federal Register. Decision time review periods are specified time limits for the agency to grant or deny pesticide registrations. PRIA also requires the Office of Inspector General (OIG) to perform an analysis of the agency's compliance with decision time review periods. The agency was in compliance with the statutory decision timeframes.

As part of obtaining a reasonable assurance as to whether the agency's financial statements are free of material misstatement, we tested compliance with those laws and regulations that could either materially affect the PRIA financial statements or that we considered significant to the audit. The objective of our audit, including our tests of compliance with applicable laws and regulations, was not to provide an opinion on overall compliance with such provisions. Accordingly, we do not express such an opinion. We did not identify any noncompliances that would result in a material misstatement to the audited financial statements.

Management's Discussion and Analysis Section of the Financial Statements

Our audit work related to the information presented in the Management's Discussion and Analysis of the pesticide program included comparing the overview information with information in the EPA's principal financial statements for consistency. We did not identify any material inconsistencies between the information presented in the two documents.

Prior Audit Coverage

During previous financial statement audits, we reported the following significant deficiencies. EPA materially understated the PRIA payroll and benefits payable, and related payroll expenses included in gross costs, in FY 2011. The agency's practice of transferring employees and expenses and liabilities from PRIA to the EPM Fund for cash flow reasons led to the understatement. The agency did not record accounts receivable for a PRIA fee until the payments were 18 months overdue. The finance center was unable to record an allowance because there was no accounting model for a PRIA allowance for doubtful accounts.

The agency has taken action to correct the deficiencies by correcting the FY 2011 payroll and benefits payable amounts in the PRIA Fund financial statements. The agency has established general ledger posting models in Compass for PRIA allowances and possible write-offs as well as policies and procedures that identify when receivables should be recorded for nonpayment of PRIA fees.

Agency Comments and OIG Evaluation

In a memorandum dated November 19, 2013, the agency responded to our draft report. The agency agreed with our findings and recommendations. The agency's complete response is included as appendix B to this report.

Paul C. Curtis
Director, Financial Statement Audits
Office of Inspector General
U.S. Environmental Protection Agency
December 17, 2013

Material Weaknesses

Table of Contents

1 – EPA Materially Overstated Expenses From Other Appropriations That Support PRIA

In its draft financial statements for FY 2012 and financial statements for FY 2011, the EPA materially overstated the expenses from other appropriations that support the PRIA Fund. This occurred because the agency does not have an effective or efficient system to accurately accumulate and report the costs incurred by other appropriations in support of PRIA Fund activities. This overstatement of the expenses from other appropriations resulted in a material overstatement of the total costs of the PRIA Fund by $14.1 million in FY 2012 and $1.7 million in FY 2011

The U.S. Government Accountability Office's *Standards for Internal Control in the Federal Government* require accurate and timely recording of transactions and events. The FMFIA emphasizes the need for agencies to provide reasonable assurance that accounts are properly recorded and accounted for to ensure reliability of financial reporting.

PRIA activities are funded by the collection of service fees from pesticides manufacturers which supplement the Office of Pesticide Programs' (OPP's) appropriated funds. Our audit work on the FY 2012 PRIA Fund financial statements showed that all of OPP's FYs 2012 and 2011 EPM expenses were being charged to either PRIA or the Pesticides Reregistration and Expedited Processing Fund (known as the FIFRA Fund). It is incorrect to charge all of OPP's EPM expenses to FIFRA and PRIA because OPP uses its EPM funds for all of its activities and not just for activities related to FIFRA or PRIA. After we identified this error, the Office of the Chief Financial Officer (OCFO) worked with OPP to compute the correct amount of FYs 2012 and 2011 expenses from other appropriations that supported the PRIA Fund. The improper charging of OPP's EPM expenses to PRIA resulted in the total costs of the PRIA program being overstated by $14.1 million in FY 2012 and $1.7 million in FY 2011. This material error caused the agency to restate the FY 2011 financial statements to reflect the decrease of the expenses from other appropriations that support the PRIA Fund and to make corresponding adjustments to the other related accounts.

Historically, the OCFO has been producing the PRIA financial statements based solely upon information contained in the EPA's accounting system. However, the EPA's accounting system does not contain sufficiently detailed information to accurately identify OPP's other appropriated expenses that relate solely to the PRIA Fund activities. While OPP does know what expenses from other appropriations support PRIA activities, it has not developed an effective and efficient method to accumulate and report these costs. The method that OPP and OCFO recently developed to revise and correct the FYs 2012 and 2011 amount of expenses from other appropriations that support PRIA relies heavily on manual computations. These manual computations are inefficient and prone to error.

Another contributing factor to the error which resulted in the material overstatement of the expenses from other appropriations is the lack of involvement by OPP in the financial statement preparation process. While the OCFO prepares the PRIA Fund financial statements, it does not have extensive knowledge of OPP business operations. Until we identified the error in the FYs 2012 and 2011 expenses from other appropriations that support PRIA, OPP had not reviewed the draft FY 2012 PRIA Fund financial statements. If OPP had carefully reviewed the draft financial statements prior to its submission to the OIG, this material error may have been avoided.

Recommendations

We recommend that the Office of the Chief Financial Officer:

1. Correct the PRIA financial statements to reflect the proper expenses paid by other appropriations.

2. Ask OPP to carefully review and comment on the draft and final versions of the PRIA Fund financial statements prior to their submission to the OIG.

We recommend that the Office of Chemical Safety and Pollution Prevention:

3. In consultation with the OCFO and other subject matter experts, develop a process that will provide accurate and timely allocation of EPM expenses from other appropriations that support the PRIA Fund.

Agency Response and OIG Evaluation

The agency agreed with our findings and recommendations, and has completed corrective actions on recommendation 1.

Agency actions on recommendation 2 are pending. OCFO will request the Office of Chemical Safety and Pollution Prevention to carefully review and comment on the draft and final versions of the PRIA Fund financial statements prior to their submission to the OIG. The estimated completion date for this corrective action is March 28, 2014.

Agency actions on recommendation 3 are pending. The Office of Chemical Safety and Pollution Prevention, in consultation with OCFO and other subject matter experts, will develop a process to ensure accurate allocations of expenses from other appropriations that support the PRIA Fund. The estimated completion date for this corrective action is December 31, 2014.

The agency's complete response is included in appendix B to this report. We agree with the agency's proposed corrective actions and believe they adequately address the issues raised.

2 – EPA Understated PRIA Payroll Liabilities
Covered by Budgetary Resources

In its draft financial statements for FY 2012, EPA materially understated the PRIA fund payroll liabilities covered by budgetary resources, as well as related payroll expense included in gross costs. OMB Circular A-136, *Financial Reporting Requirements,* and Statement of Federal Financial Accounting Standards No. 5 require that liabilities be recognized when they are incurred. The agency's practice of transferring employees and related expenses and liabilities from PRIA to the EPM Fund for cash flow reasons led to the understatement. The FY 2011 payroll liabilities covered by budgetary resources for PRIA was $500,000, while the FY 2012 payroll liabilities covered by budgetary resources was zero. Such understatements could impact the opinion on the financial statements and the reliance on reported PRIA financial information. This understatement is a recurring issue which needs resolution.

Statement of Federal Financial Accounting Standards No. 5 states that liabilities should be recognized for exchange transaction, such as when a federal employee performs services in exchange for compensation, when the services are provided. OMB Circular No. 136 states: "Liabilities shall be recognized when they are incurred regardless of whether they are covered by available budgetary resources."

OPP transferred all employees from PRIA to EPM at the end of FY 2012 pay period 13. EPA uses EPM for a broad range of abatement, prevention and compliance activities and personnel compensation, benefits, travel and expenses for all programs of the agency. On average, 58 employees were assigned PRIA throughout FY 2012. The transfer removed the base upon which the payroll liabilities covered by budgetary resources are calculated. As a result, payroll liabilities covered by budgetary resources were significantly understated.

EPA began the practice of moving payroll expenses from PRIA to EPM in FY 2000. When PRIA resources are low, the agency transfers employees from PRIA to EPM to keep PRIA obligations and disbursements within budgetary and cash limits. As PRIA fees are collected, employees are moved back to the PRIA appropriation. EPA disclosed this ongoing practice in prior PRIA financial statement reports, and this practice is expected to continue throughout FY 2013. Temporarily moving employees for cash flow reasons should not impact accruals as long as those employees are continuing the same work. If the transfers become permanent, PRIA should recognize a benefit since another appropriation would be covering the accrued payroll debt.

The process of moving employees and related payroll expenses and liabilities from PRIA to EPM contributed to the understatement of the PRIA payroll liabilities in the draft FY 2012 financial statements. However, the OCFO should have realized that the transfer of employees from PRIA to EMP was only temporary and computed payroll liability amounts accordingly.

Recommendations

We recommend that the Office of the Chief Financial Officer:

4. Correct the PRIA financial statements to reflect the proper payroll liability amounts.

5. Closely monitor the payroll liability amounts for PRIA at year-end.

Agency Response and OIG Evaluation

The agency agreed with our findings and recommendation, and has completed corrective actions on recommendation 4. OCFO corrected the PRIA financial statements to reflect the proper payroll liability amounts.

Agency actions on recommendation 5 are pending. The estimated completion date for this corrective action is September 30, 2014.

The agency's complete response is included in appendix B to this report.

Status of Recommendations and Potential Monetary Benefits

Rec. No.	Page No.	Subject	Status[1]	Action Official	Planned Completion Date	Claimed Amount	Agreed-To Amount
1	8	Correct the PRIA financial statements to reflect the proper expenses paid by other appropriations.	C	Office of the Chief Financial Officer	09/24/13		
2	8	Ask OPP to carefully review and comment on the draft and final versions of the PRIA Fund financial statements prior to their submission to the OIG.	O	Office of the Chief Financial Officer	03/28/14		
3	8	In consultation with the OCFO and other subject matter experts, develop a process that will provide accurate and timely allocation of EPM expenses from other appropriations that support the PRIA Fund.	O	Office of Chemical Safely and Pollution Prevention	12/31/14		
4	10	Correct the PRIA financial statements to reflect the proper payroll liability amounts.	C	Office of the Chief Financial Officer	02/26/13		
5	10	Closely monitor the payroll liability amounts for PRIA at year-end.	O	Office of the Chief Financial Officer	09/30/14		

[1] O = recommendation is open with agreed-to corrective actions pending
C = recommendation is closed with all agreed-to actions completed
U = recommendation is unresolved with resolution efforts in progres

FYs 2012 and 2011 (RESTATED) PESTICIDE REGISTRATION FUND (PRIA) FINANCIAL STATEMENTS

Produced by the U.S. Environmental Protection Agency
Office of the Chief Financial Officer
Office of Financial Management

14-1-0042

TABLE OF CONTENTS

Management's Discussion and Analysis

MANAGEMENT'S DISCUSSION AND ANALYSIS

The Agency's Office of Pesticide Programs (OPP) was established to administer the Federal Insecticide, Fungicide and Rodenticide Act (FIFRA) to protect public health and the environment. The law requires the Agency to balance public health and environmental concerns with the expected economic benefits derived from pesticides. The guiding principles of the pesticide program are to reduce risks from pesticides in food, the workplace, and other exposure pathways and to prevent pollution by encouraging the use of new and safer pesticides.

With passage of the Pesticide Registration Improvement Act (PRIA) of 2003, the pesticide program now administers the Pesticide Registration Fund. PRIA authorizes the collection of new fees for pesticide registrations. Registration service fees are deposited into the Registration Fund and made available for obligation to the extent provided in appropriation Acts, and are available without fiscal year limitation.

Pesticide Registration

Under the authority of FIFRA and the Federal Food, Drug, and Cosmetic Act (FFDCA) as amended by the Food Quality Protection Act (FQPA), no person or State can distribute or sell any pesticide that is not registered with the Agency. The pesticide registration program works to decrease the risk to the public from pesticide use through the regulatory review of new pesticides. In 2004, Congress passed PRIA 1, with deadlines for completion of certain registration actions. As part of the registration program, EPA expedites the registration of reduced-risk pesticide uses, which are generally presumed to pose lower risks to consumers, workers, groundwater, and/or wildlife. These accelerated pesticide reviews provide an incentive for industry to develop, register, and use lower risk pesticides. Additionally, the availability of these reduced-risk pesticides provides alternatives to older, potentially more harmful products currently on the market.

Biological agents are potential weapons that could be exploited by terrorists against the United States. EPA's pesticides antimicrobial program is working to help address this threat. Antimicrobials play an important role in public health and safety. EPA is conducting comprehensive scientific assessments and developing test protocols to determine the safety and efficacy of products used against chemical and biological weapons of mass destruction, and registering products as necessary. EPA is also developing a timeline for prioritizing and implementing the tests. In addition, the Section 18 program provides emergency exemption to any part of FIFRA. This authority is typically used by States on an emergency basis. EPA has recently used this authority to help with homeland security. Section 18 exemptions have been authorized to help with anthrax and soybean rust.

PRIA established registration service fees for certain antimicrobials, biopesticides and conventional pesticides registration actions. The category of action, the amount of the registration service fee, and the corresponding decision review periods by year are prescribed in the statute. The goal is to create a more predictable evaluation process for affected pesticide decisions, and couple the collection of individual fees with specific decision review periods. The legislation also promotes shorter decision review periods for reduced-risk applications. PRIA 1 became effective on March 23, 2004, and the collection of registration fees were authorized through FY 2008. PRIA 1 was reauthorized with passage of the Pesticide Registration Improvement Renewal Act (commonly referred to as PRIA 2) on October 9, 2007. PRIA 2 became effective retroactive to October 1, 2007, and the collection of registration fees were authorized through FY 2012. PRIA 2 was reauthorized with the passage of the Pesticide Registration Improvement Extension Act (referred to as PRIA 3) on September 28, 2012 and became effective 2 days later on October 1, 2012.

In order for a pending or a new application covered by PRIA to be deemed complete and subject to the decision review periods, a registrant is required to pay the applicable fee or receive a waiver from the fees. For most applications, the decision review period starts 21 days after submission of the application - provided that the fee has been paid, fee waiver granted or in the case of a 75% or 50% fee waiver under PRIA 3, the fee has been paid and waiver granted. The legislation provides fee waivers for certain categories of small businesses, and minor uses[1]. Exemption from the requirement to pay a registration service fee is continued under PRIA 3 for applications solely associated with IR-4 petitions[2]. Applications from federal and state agencies are also exempt from registration service fees. If the registrant requests a waiver or reduction of the fee, the decision review period will begin when the Agency grants such request or in the case of small business fee waivers, no more than 60 days after receipt of the waiver application. If it is determined that a fee is required and thus the waiver is not granted, the decision review period starts after the fee is collected.

Applications received prior to October 1, 2007 were covered by PRIA 1. Applications received up to September 30, 2012 were covered by PRIA 2[3] and applications received on or after October 1, 2012 are covered by PRIA 3. PRIA 3 contains the same audit provision as PRIA 2. PRIA 3 provides new authority to reject an application if an application fails a preliminary technical screen; PRIA 3 increases the fee categories or types of applications covered by PRIA from 140 to 189, and maintains set asides to support worker protection and applicator training activities as well as IPM grants at comparable levels to PRIA 2.

Research Program Description

EPA's Chemical Safety for Sustainability (CSS) research program is leading the sustainable development, use, and assessment of chemicals and materials by advancing integrated chemical evaluation strategies and decision support tools that promote human and environmental health

---

[1] Minor use pesticides are those that produce relatively little revenue for their manufacturers, for a variety of reasons. They may be registered for a seldom seen pest, or for a crop that is not grown by a large number of producers. However, minor crops include some high revenue fruit, vegetable, and ornamental crops.
[2] The IR-4 (Interregional Research Project No.4) program is involved in making sure that pesticides are registered for use on minor crops. IR-4 helps by conducting research on minor use pesticides, pesticides that would not otherwise be profitable to manufacture.

[3] Out of approximately 8,186 completed PRIA 2 actions more than 99% were completed on or before the PRIA 2 due date.

and are protective of vulnerable species and populations. The research is focused on providing integrated solutions in support of the Agency's efforts to manage chemical (including pesticides and toxics) risks. The data, methods and tools developed will guide the prioritization and testing process, from screening approaches through more complex testing and assessments. The research program's major goals are: (1) to build the knowledge infrastructure to support scientific discovery and sustainable decisions, (2) to develop and apply rapid, efficient, and effective methods for improved chemical prioritization, screening, and testing, (3) to provide models and tools necessary to make decisions supporting safe use across the chemical lifecycle.

Current testing and assessment approaches are resource intensive and lack data sufficient to meet decision-making needs posed by the large and growing number of chemicals. The CSS ToxCast Program performs cost-effective, state-of-the-art chemical screening to assess how chemicals may affect human health. ToxCast simultaneously tests thousands of chemicals using hundreds of high-throughput and high-content approaches. This allows the EPA to directly examine environmental chemicals' role in human disease processes, cell systems, and pathway targets. The ToxCast program has moved beyond the proof-of-concept phase focus on pesticide actives. Results of Phase II of this program, which covers 1,860 chemicals, will be released and publicly available in FY13.

In providing research on methods, models, and data to support decision-making regarding specific individual or classes of pesticides and toxic substances that are of high priority, the program will continue to develop:
- Predictive biomarkers, quantitative structure activity relationships, and alternative test methods for prioritizing and screening chemicals for a number of adverse effects (e.g., neurotoxicity, reproductive toxicity) that will lead to a reduction in and more efficient use of whole animals in toxicity testing; and
- Approaches for applying high-throughput screening and computational models developed under the ToxCast program to support prioritization of chemicals for further testing under EPA's Endocrine Disruptor Screening Program.
- Data and protocols on the impact of waste water treatment technologies on pesticides and their products of transformation.

To support the development of probabilistic risk assessments to protect endangered populations of birds, fish, other wildlife, and non-target plants from pesticides while making sure farmers and communities have the pest control tools they need, this program has four key research components:
- Extrapolation among wildlife species and exposure scenarios of concern;
- Population biology to improve population dynamics in spatially-explicit habitats;
- Models for assessing the relative risk of chemical and non-chemical stressors; and
- Models to define geographical regional/spatial scales for risk assessment.

Methods for characterization of population-level risks of toxic substances to aquatic life and wildlife also are being developed as part of the Agency's long-term goal of developing scientifically valid approaches for assessing spatially-explicit, population-level risks to wildlife populations and non-target plants and plant communities from pesticides, toxic chemicals and multiple stressors while advancing the development of probabilistic risk assessment.

The program anticipates that the Agency will be better positioned to perform its mission of protecting human health and the environment as scientific information becomes digitized and readily available, methods and models to capture the complexities of chemical exposure and hazard in toxicity testing are developed and approaches focused on development of more sustainable alternatives are provided to decision-makers.

Enforcement and Compliance Assurance Program Description

The Pesticide Enforcement and Compliance Assurance Program focuses on pesticide product and user compliance. These include problems relating to pesticide worker safety, certification and training of applicators, ineffective antimicrobial products, food safety, adverse effects, risks of pesticides to endangered species, pesticide containers and containment facilities, and e-commerce and misuse. The enforcement and compliance assurance program provides compliance assistance to the regulated community through its National Agriculture Compliance Assistance Center, seminars, guidance documents, brochures, and other forms of communication to ensure knowledge of and compliance with environmental laws.

EPA's grant support to states' and tribes' pesticide programs emphasizes its commitment to maintaining a strong compliance and enforcement presence. Agency Cooperative Agreement priorities for FY2008 - FY2010 include the enforcement of worker protection standards; compliance monitoring and enforcement activities related to the newly promulgated pesticide container and containment rules, and program performance reporting. Core program activities include inspections of producing establishments; dealers/distributors/retailers; e-commerce; imports and exports, and pesticide misuse. Additionally, through the Cooperative Agreement resources we support inspector training and training for state/tribal senior managers, scientists, and supervisors.

Highlights and Accomplishments

Registration Financial Perspective

During FY 2012, the Agency's obligations charged against the PRIA Fund for the cost of registration were $13.1 million and 53.4 workyears. Of this amount, OPP obligated $7.4M in PC&B.

Appropriated funds are used in addition to Registration funds. In FY 2012, the Enacted Operating Plan included approximately $ 47.2 million in appropriated funds for registration activities. The unobligated balance in the Fund at the end of FY 2012 was $6.8 million.

The Fund has two types of receipts: fee collections and interest earned on investments. Of the $15.6 million in FY 2012 receipts, more than 99.9% were fee collections.

Registration Program Performance Measures

The following measures support the program's strategic goals of Healthy Communities and Ecosystems as contained in the FY 2012 President's budget.

Measure 1: Number of new active ingredients registered.

Results: In FY2012 EPA registered 35 new active ingredients, of which 21 are biopesticides, 11 are conventional pesticides (including one new active ingredient with import tolerance use only) and 3 are antimicrobial pesticides. This measure includes both reduced-risk and non-reduced-risk pesticides.

Measure 2: Progress in Registering Reduced-risk Pesticides.

Results: In FY 2012, EPA registered 23 reduced-risk new active ingredients, 21 of which were biological pesticides and 2 of which were conventional pesticides. Biological pesticides are certain types of pesticides derived from such natural materials as animals, plants, bacteria, and certain minerals. They are usually less toxic and are typically considered safer pesticides than the traditional conventional chemicals; therefore, the 21 biopesticides new active ingredients are counted as reduced-risk pesticides. Conventional "reduced risk" pesticides have one or more of the following advantages over currently registered pesticides: low impact on human health, low toxicity to non-target organisms, low potential for groundwater contamination, lower use rates, low pest resistance potential, and compatibility with integrated pest management strategies.

Measure 3: Number of New Food Uses Registered.

Results: EPA registered 161 new uses for previously registered active ingredients. Of these new uses, 158 were for conventional pesticides, 2 were for antimicrobial pesticides, and 1 was for a biopesticide

Measure 4: Progress in Registering Reduced-risk New Uses.

Results: Included in the new uses registered are 10 reduced-risk uses, of which 7 were associated with conventional pesticides and 3 were biopesticide new uses.

PRINCIPAL
FINANCIAL STATEMENTS

TABLE OF CONTENTS

Financial Statements

Notes to Financial Statements

Environmental Protection Agency
PRIA
Balance Sheet
For the Years Ended September 30, 2012 and 2011
(Dollars in Thousands)

	FY 2012	FY 2011
ASSETS		
Intragovernmental:		
Fund Balance With Treasury (Note 2)	$ 12,443	$ 11,241
Other (Note 3)	-	40
Total Intragovernmental	$ 12,443	$ 11,281
Accounts Receivable, Net (Note 5)	-	2
Property, Plant & Equipment, Net (Note 4)	2,753	3,188
Total Assets	$ **15,196**	$ **14,471**
LIABILITIES		
Intragovernmental:		
Accounts Payable and Accrued Liabilities	93	133
Other (Note 5)	74	95
Total Intragovernmental	$ 167	$ 228
Accounts Payable & Accrued Liabilities	$ 757	$ 816
Payroll & Benefits Payable (Note 6)	2,022	962
Other (Note 5)	11,277	10,064
Total Liabilities	$ 14,223	$ 12,070
NET POSITION		
Cumulative Results of Operations	973	2,401
Total Net Position	973	2,401
Total Liabilities and Net Position	$ **15,196**	$ **14,471**

The accompanying notes are an integral part of these statements.

Environmental Protection Agency
PRIA
Statement of Net Cost
For the Years Ended September 30, 2012 and 2011
(Dollars in Thousands)

	FY 2012	Restated FY 2011
COSTS		
Gross Costs (Note 9)	$ 15,848	$ 17,672
Expenses from Other Appropriations (Note 7)	29,726	35,993
Total Costs	$ 45,574	$ 53,665
Less:		
Earned Revenue (Notes 8 and 9)	14,396	15,809
NET COST OF OPERATIONS (Note 9)	$ 31,178	$ 37,856

The accompanying notes are an integral part of these statements.

Environmental Protection Agency
PRIA
Statement of Changes in Net Position
For the Years Ended September 30, 2012 and 2011
(Dollars in Thousands)

		FY 2012		Restated FY 2011
Cumulative Results of Operations:				
Net Position - Beginning of Period		2,401		4,064
Beginning Balances, as Adjusted	$	2,401	$	4,064
Budgetary Financing Sources:				
Nonexchange Revenue - Securities Investment		1		5
Nonexchange Revenue - Other		12		-
Income from Other Appropriations (Note 7)		29,726		35,993
Total Budgetary Financing Sources	$	29,739	$	35,998
Other Financing Sources (Non-Exchange)				
Imputed Financing Sources		11		195
Total Other Financing Sources	$	11	$	195
Net Cost of Operations		(31,178)		(37,856)
Net Change		(1,428)		(1,663)
Cumulative Results of Operations	$	973	$	2,401

The accompanying notes are an integral part of these statements.

Environmental Protection Agency
PRIA
Statement of Budgetary Resources
For the Years Ended September 30, 2012 and 2011
(Dollars in Thousands)

	FY 2012	FY 2011
BUDGETARY RESOURCES		
Unobligated Balance, Brought Forward, October 1:	$ 4,247	$ 7,393
Unobligated balance brought forward, October 1, as adjusted	4,247	7,393
Recoveries of Prior Year Unpaid Obligations	43	-
Other changes in unobligated balance	-	(40)
Unobligated balance from prior year budget authority, net	4,290	7,353
Appropriations (discretionary and mandatory)	15,619	11,790
Spending authority from offsetting collections (discretionary and mandatory)	39	-
Total Budgetary Resources	$ 19,948	$ 19,143
STATUS OF BUDGETARY RESOURCES		
Obligations incurred	$ 13,192	$ 14,896
Unobligated balance, end of year:		
Apportioned	6,756	4,247
Total unobligated balance, end of period	6,756	4,247
Total Status of Budgetary Resources	$ 19,948	$ 19,143
CHANGE IN OBLIGATED BALANCE		
Unpaid Obligations, Brought Forward, October 1 (gross)	$ 6,955	$ 7,701
Obligated balance, start of year (net), before adjustments	6,955	- 7,701
Obligated balance, start of year (net), as adjusted	6,955	7,701
Obligations incurred	13,192	14,896
Outlays (gross)	(14,460)	(15,642)
Recoveries of prior year unpaid obligations	(43)	-
Obligated balance, end of period		
Unpaid obligations, end of year (gross)	5,644	6,955
Obligated balance, end of period (net)	$ 5,644	$ 6,955
BUDGET AUTHORITY AND OUTLAYS, NET:		
Budget authority, gross (discretionary and mandatory)	$ 15,658	$ 11,790
Actual offsetting collections (discretionary and mandatory)	(39)	-
Budget authority, net (discretionary and mandatory)	$ 15,619	$ 11,790
Outlays, gross (discretionary and mandatory)	$ 14,460	$ 15,642
Actual offsetting collections (discretionary and mandatory)	(39)	-
Outlays, net (discretionary and mandatory)	14,421	15,642
Distributed offsetting receipts	(15,622)	(11,790)
Agency outlays, net (discretionary and mandatory)	$ (1,201)	$ 3,852

The accompanying notes are an integral part of these statements.

Environmental Protection Agency
PRIA
Notes to Financial Statements
(Dollars in Thousands)

Note 1. *Summary of Significant Accounting Policies*

A. Reporting Entity

The U.S. Environmental Protection Agency (EPA or Agency) was created in 1970 by executive reorganization from various components of other Federal agencies in order to better marshal and coordinate federal pollution control efforts. The Agency is generally organized around the media and substances it regulates -- air, water, land, hazardous waste, pesticides and toxic substances.

The Pesticide Registration Fund (PRIA) is authorized under the Pesticide Registration Improvement Act of 2003 (which amended the Federal Insecticide, Fungicide, and Rodenticide Act (FIFRA)), and became effective on March 23, 2004. This Act authorizes the EPA to assess and collect pesticide registration service fees on applications submitted to register pesticides covered by this Act, as well as assess and collect fees to register new active ingredients not listed in the Registration Division 2003 Work Plan of the Office of Pesticide Programs. The Pesticide Registration Improvement Renewal Act (commonly referred to as PRIA II) extended the authority to collect pesticide registration service fees through FY 2012. PRIA II became effective October 1, 2007. PRIA II was reauthorized with the passage of the Pesticide Registration Improvement Extension Act (referred to as PRIA III) on September 28, 2012 and became effective 2 days later on October 1, 2012. The PRIA Fund is accounted for under Treasury symbol number 68X5374.

The PRIA fund may charge some administrative costs directly to the fund and charge the remainder of the administrative costs to Agency-wide appropriations. Costs funded by Agency-wide appropriations for FYs 2012 and 2011 were $29,726 thousand and $35,993 thousand, respectively. This amount was included as Income from Other Appropriations on the Statement of Changes in Net Position and as Expenses from Other Appropriations on the Statement of Net Cost for FYs 2012 and 2011.

B. Basis of Presentation

These financial statements have been prepared to report the financial position and results of operations of the EPA for the Pesticide Registration Fund (PRIA) as required by the Chief Financial Officers Act of 1990 and the Pesticide Registration Improvement Act (PRIA) of 2003. In the prior years, pesticide registration was included in the FIFRA financial statements. The reports have been prepared from the books and records of the EPA in accordance with Office of Management and Budget (OMB) Circular A-136 *Financial Reporting Requirements*, and the EPA's accounting policies which are summarized in this note. These statements are therefore different from the financial reports also prepared by the EPA pursuant to OMB directives that are used to monitor and control the EPA's use of budgetary resources. The balances in these reports have been updated from the EPA consolidated financial statements to reflect the use of FY 2012 cost factors for calculating imputed costs for Federal civilian benefits programs. These updates impact the Balance Sheet, Statement of Net Cost, and Statement of Changes in Net Position.

C. Budgets and Budgetary Accounting

Funding for PRIA is provided by fees collected from industry to offset costs incurred by EPA in carrying out these programs. Each year the EPA submits an apportionment request to OMB based on the anticipated collections of industry fees.

D. Basis of Accounting

Generally Accepted Accounting Principles (GAAP) for Federal entities is the standard prescribed by the Federal Accounting Standards Advisory Board (FASAB), which is the official standard setting body for the federal government. The financial statements are prepared in accordance with GAAP for federal entities.

Transactions are recorded on an accrual accounting basis and a budgetary basis. Under the accrual method, revenues are recognized when earned and expenses are recognized when a liability is incurred, without regard to receipt or payment of cash. Budgetary accounting facilitates compliance with legal constraints and controls over the use of Federal funds. All interfund balances and transactions have been eliminated.

E. Revenues and Other Financing Sources

For FYs 2012 and 2011, PRIA received funding from fees collected from registrants requesting pesticide registrations. For FYs 2012 and 2011, revenues were recognized from fee collections to the extent that expenses are incurred during the fiscal year.

F. Funds with the Treasury

The PRIA fund deposits receipts and processes disbursements through its operating account maintained at the U.S. Department of the Treasury.

G. Investments in U. S. Government Securities

Investments in U. S. government securities are maintained by Treasury and are reported at amortized cost net of unamortized discounts. Discounts are amortized over the term of the investments and reported as interest income. PRIA holds the investments to maturity, unless needed to finance operations of the fund. No provision is made for unrealized gains or losses on these securities because, in the majority of cases, they are held to maturity.

H. General Property, Plant and Equipment

Purchases of the EPA-held personal equipment are capitalized if the equipment is valued at $25 thousand or more and has an estimated useful life of at least two years. Depreciation is taken on a basic straight-line method over the specific asset's useful life, ranging from two to15 years.

The EPA shows property, plant and equipment at net of depreciation on its audited financial statements.

All funds (except for the Working Capital Fund) capitalize software if those investments are considered Capital Planning and Investment Control (CPIC) or CPIC Lite systems with the provisions of SFFAS No. 10, "Accounting for Internal Use Software." Once software enters the production life cycle phase, it is depreciated using the straight-line method over the specific asset's useful life ranging from two to 10 years.

I. Liabilities

Liabilities represent the amount of monies or other resources that are likely to be paid by the Agency as the result of an Agency transaction or event that has already occurred and can be reasonably estimated. However, no liability can be paid by the Agency without an appropriation or other collections. Liabilities for which an appropriation has not been enacted are classified as unfunded liabilities and there is no certainty that the appropriations will be enacted. For PRIA, liabilities are liquidated from fee receipts, since PRIA receives no appropriation. Liabilities of the Agency arising from anything other than contracts can be abrogated by the Government acting in its sovereign capacity.

J. Accrued Unfunded Annual Leave

Annual, sick and other leave is expensed as taken during the fiscal year. Sick leave earned but not taken is not accrued as a liability. Annual leave earned but not taken as of the end of the fiscal year is accrued as an unfunded liability. Accrued unfunded annual leave is included in the Balance Sheet as a component of "Payroll and Benefits Payable."

K. Retirement Plan

There are two primary retirement systems for Federal employees. Employees hired prior to January 1, 1987, may participate in the Civil Service Retirement System (CSRS). On January 1, 1984, the Federal Employees Retirement System (FERS) went into effect pursuant to Public Law 99-335. Most employees hired after December 31, 1983, are automatically covered by FERS and Social Security. Employees hired prior to January 1, 1984, elected to either join FERS and Social Security or remain in CSRS. A primary feature of FERS is that it offers a savings plan to which the Agency automatically contributes one percent of pay and matches any employee contributions up to an additional four percent of pay. The Agency also contributes the employer's matching share for Social Security.

With the issuance of SFFAS No. 5, "Accounting for Liabilities of the Federal Government," accounting and reporting standards were established for liabilities relating to the federal employee benefit programs (Retirement, Health Benefits, and Life Insurance). SFFAS No. 5 requires that the employing agencies recognize the cost of pensions and other retirement benefits during their employees' active years of service. SFFAS No. 5 requires that the Office of Personnel Management (OPM), as administrator of the CSRS and FERS, the Federal Employees

Health Benefits Program, and the Federal Employees Group Life Insurance Program, provide federal agencies with the actuarial cost factors to compute the liability for each program.

L. Offsetting Receipts

Beginning in FY 2007 OMB Circular A-136, *Financial Reporting Requirements,* requires that the amount of distributed offsetting receipts reported in the Statement of Budgetary Resources (SBR) should equal the amount recorded as offsetting receipts by the Department of the Treasury (Treasury). Pesticide Registration Fees collected under PRIA are considered to be offsetting receipts by Treasury.

M. Use of Estimates

The preparation of financial statements requires management to make certain estimates and assumptions that affect the reported amounts of assets and liabilities and the reported amounts of revenue and expenses during the reporting period. Actual results could differ from those estimates.

Note 2. Fund Balance with Treasury

		FY 2012	FY 2011
Revolving Funds:	Entity Assets	$ **12,443**	$ **11,241**

Note 3. Other Assets

Other Assets consist of advances for Interagency Agreements. As of September 30, 2012 and 2011, funds advanced that will be applied to future costs as incurred were $0 and $40 thousand respectively.

Note 4. General Property, Plant and Equipment

General property, plant and equipment consists of the EPA-Held personal property, software, and software in development.

As of September 30, 2012 and 2011, General Property, Plant and Equipment consist of the following:

	FY 2012			FY 2011		
	Acquisition Value	Accumulated Depreciation	Net Book Value	Acquisition Value	Accumulated Depreciation	Net Book Value
EPA-Held Equipment	$ 410	$ (305)	$ 105	$ 410	$ (271)	$ 139
Software	4,458	(1,810)	2,648	4,198	(1,149)	3,049
Total	$ 4,868	$ (2,115)	$ 2,753	$ 4,608	$ (1,420)	$ 3,188

Note 5. Other Liabilities

For FYs 2012 and 2011, Payroll and Benefits Payable, non-federal, are presented on a separate line of the Balance Sheet and in a separate footnote (see Note 6).

	FY 2012	FY 2011
Other Intragovernmental Liabilities - Covered by Budgetary Resources		
Employer Contributions - Payroll	$ 74	$ 95
Total	$ 74	$ 95
Other Non-Federal Liabilities - Covered by Budgetary Resources		
Advances from Non-Federal Entities	$ 11,277	$ 10,064
Total	$ 11,277	$ 10,064

Note 6. Payroll and Benefits Payable, Non-Federal:

	FY 2012	FY 2011
Covered by Budgetary Resources		
Accrued Payroll Payable to Employees	$ 415	$ 327
Withholdings Payable	28	61
Thrift Savings Plan Benefits Payable	18	17
Total	**$ 461**	**$ 405**
Not Covered by Budgetary Resources		
Unfunded Annual Leave	$ 1,561	$ 557
Total	**$ 1,561**	**$ 557**

At various periods throughout FYs 2012 and 2011 employees with their associated payroll costs were transferred from PRIA to the Environmental Programs and Management (EPM) appropriation. (See graph in Note 7 below showing trend of hours charged per month to the PRIA fund for FYs 2012 and 2011.) These employees were transferred in order to keep PRIA's obligations and disbursements within budgetary limits.

This process has led to variations between the year-end liabilities of FYs 2012 and 2011. The liabilities covered by budgetary resources (both intragovernmental and non-Federal) represent unpaid payroll and benefits at year-end. For FY 2012 Pay Period 26; no employees charged any part of their salary and benefits to PRIA. As of September 30, 2012, the liabilities were $74 thousand and $461 thousand for employer contributions and accrued funded payroll and benefits as compared to FY 2011's balances of $95 thousand and $405 thousand, respectively.

In contrast, the unfunded annual leave liability is a longer term liability than the funded liabilities. At various periods throughout FYs 2012 and FY 2011, approximately 211 and 130 employees, respectively, in total have been under PRIA's accountability. As of September 30, 2012 and 2011 liability balances for unfunded annual leave were accrued to cover these employees for a total of $1.5 million and $557 thousand, respectively.

Note 7. *Income and Expenses from Other Appropriations*:

The Statement of Net Cost reports program costs that include the full costs of the program outputs and consist of the direct costs and all other costs that can be directly traced, assigned on a cause and effect basis, or reasonably allocated to program outputs.

During FYs 2012 and 2011, the EPA had two appropriations which funded a variety of programmatic and non-programmatic activities across the Agency, subject to statutory requirements. The EPM appropriation was created to fund personnel compensation and benefits, travel, procurement, and contract activities. Transfers of employees from PRIA to EPM at various times during FYs 2012 and 2011 (see Note 6 above) resulted in an increase in payroll expenses in EPM, and these costs financed by EPM are reflected as an increase in the Expenses from Other Appropriations on the Statement of Net Cost. The increased financing from EPM is reported on the Statement of Changes in Net Position as Income from Other Appropriations.

In terms of hours charged to PRIA each month, the transfers of employees and their associated costs during FYs 2012 and 2011 are shown below. Note that a decrease in hours charged to PRIA normally signifies an increase in EPM's payroll costs, and vice versa.

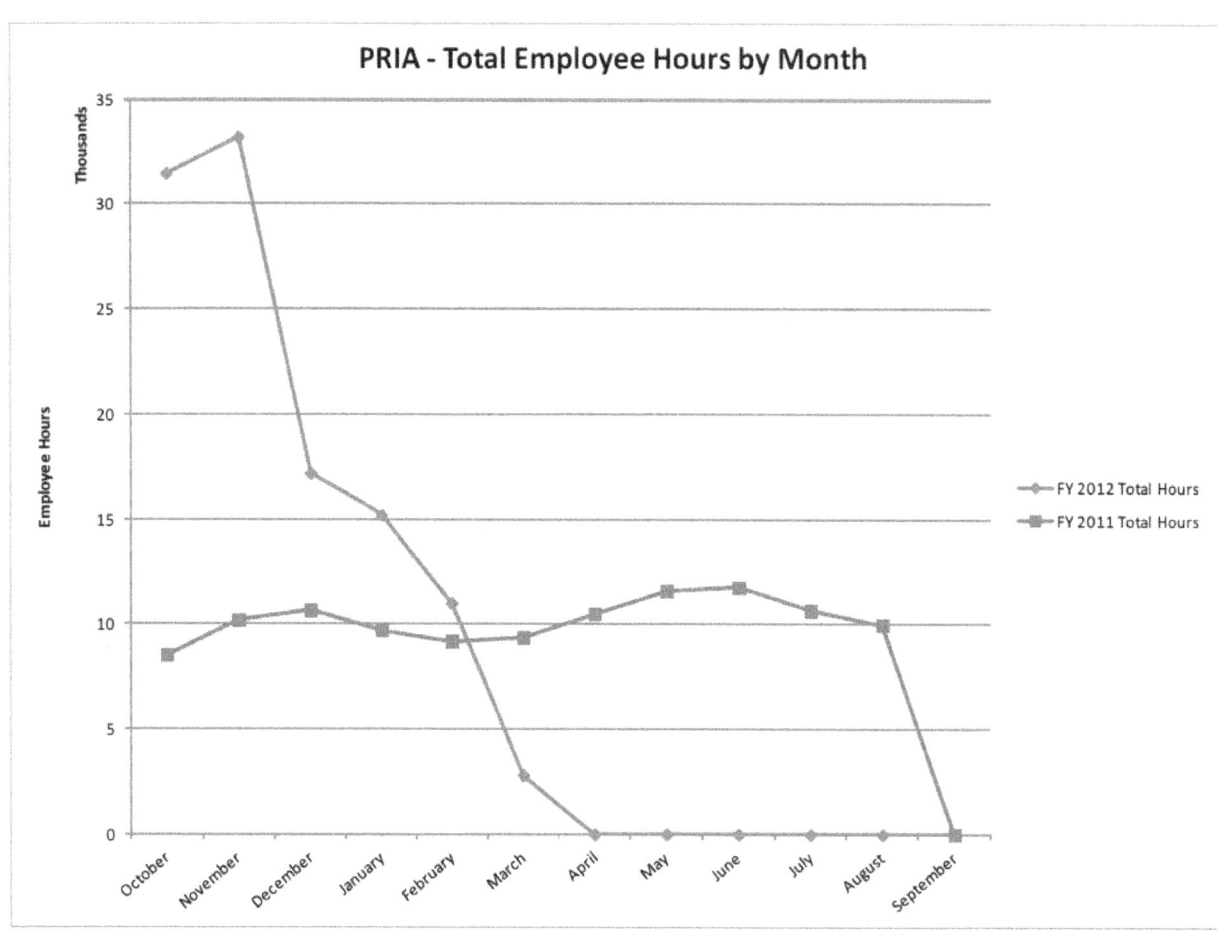

The EPM costs related to PRIA are allocated based on specific EPM program codes which have been designated for Pesticide registration activities. As illustrated below, there is no impact on PRIA's Statement of Changes in Net Position.

	Income From Other Appropriations	Expenses From Other Appropriations	Net Effect
FY 2012	$ 29,726	$ 29,726	$ 0
Restated FY 2011	$ 35,993	$ 35,993	$ 0

Note 8. *Exchange Revenues, Statement of Net Cost*

For FYs 2012 and 2011, the exchange revenues reported on the Statement of Net Cost consists of non-Federal amounts.

Note 9. *Intragovernmental Costs and Exchange Revenue*

	FY 2012		Restated FY 2011
COSTS:			
Intragovernmental	$ 2,171	$	2,661
With the Public	13,677		15,011
Expenses from Other Appropriations	29,726		35,993
Total Costs	$ 45,574	$	53,665
REVENUE:			
With the Public	14,396		15,809
Total Revenue	$ 14,396	$	15,809
NET COST OF OPERATIONS	$ **31,178**	$	**37,856**

Intragovernmental costs relate to the source of the goods or services not the classification of the related revenue.

Note 10. *Reconciliation of Net Cost of Operations to Budget (formerly the Statement of Financing)*

		FY 2012		Restated FY 2011
RESOURCES USED TO FINANCE ACTIVITIES:				
Budgetary Resources Obligated				
Obligations Incurred	$	13,192	$	14,896
Less: Spending Authority from Offsetting Collections and Recoveries		(82)		-
Obligations, Net of Offsetting Collections	$	13,110	$	14,896
Less: Offsetting Receipts (Note 1 Section L)		(12)		(11,790)
Net Obligations		13,098		3,106
Other Resources				
Imputed Financing Sources	$	11	$	195
Income from Other Appropriations (Note 7)		29,726		35,993
Net Other Resources Used to Finance Activities	$	29,737	$	36,188
Total Resources Used To Finance Activities	$	42,835	$	39,294
RESOURCES USED TO FINANCE ITEMS				
NOT PART OF NET COST OF OPERATIONS				
Change in Budgetary Resources Obligated	$	1,286	$	913
Offsetting Receipts Not Affecting Net Cost (Note 1 Section L)		12		11,790
Resources that Finance Asset Acquistion		(260)		(3,966)
Total Resources Used to Finance Items Not				
Part of the Net Cost of Operations	$	1,038	$	8,737
Total Resources Used to Finance the Net				
Cost of Operations	$	43,873	$	48,031
COMPONENTS OF NET COST OF OPERATIONS				
THAT WILL NOT REQUIRE OR GENERATE				
RESOURCES IN THE CURRENT PERIOD				
Components Requiring or Generating Resources in Future Periods:				
Increase in Annual Leave Liability	$	1,004	$	411
Increase in Public Exchange Revenue Receivable	$	(14,396)		(15,810)
Total Components of Net Cost of Operations that				
Requires or Generates Resources in the Future	$	(13,392)	$	(15,399)
Components Not Requiring/Generating Resources:				
Depreciation and Amortization		696		1,182
Expenses Not Requiring Budgetary Resources		1		4,042
Total components of Net cost of Operations that Will Not Require or Generate Resources		697		5,224
Total components of Net cost of Operations that Will Not Require				
or Generate Resources in the Current Period		(12,695)		(10,175)
Net Cost of Operations	$	**31,178**	$	**37,856**

Note 11. Restatements

EPA discovered an accounting error that resulted in the material misstatement of EPA's financial statements issued for the period FY 2011. As a consequence, EPA is correcting the errors by restating its Statement of Net Cost and Statement of Changes in Net Position as of September 30, 2011.

The effect of the restatement is as follows:

	FY 2011, as Previously Reported	Adjustment	FY 2011, as Restated
Statement of Net Cost			
Expenses from Other Appropriations (Note 6)	36,710	(717)	35,993
Total Costs	54,382	(717)	53,665
Net Cost of Operations (Note 9)	38,573	(717)	37,856
Statement of Changes in Net Position			
Income from Other Appropriations (Note 6)	36,710	(717)	35,993
Total Budgetary Financing Sources	36,715	(717)	35,998
Net Cost of Operations (Note 9)	(38,573)	717	(37,856)

Agency's Response to Draft Report

NOV 1 9 2013

MEMORANDUM

SUBJECT: Response to Office of Inspector General Draft Report No. 0A-FY13-
0080 *"Fiscal Years 2012 and 2011 (Restated) Financial Statements for
the Pesticide Reregistration Fund,"* dated November 4, 2013

FROM: Maryann Froehlich
            Acting Chief Financial Officer

            James J. Jones, Assistant Administrator
            Office of Chemical Safety and Pollution

TO: Arthur A. Elkins, Jr.
            Inspector General

Thank you for the opportunity to respond to the issues and recommendations in the subject audit
report. Following is a summary of the agency's overall position, along with its position on each
of the report recommendations. We have provided high-level intended corrective actions and
estimated completion dates to the extent we can.

AGENCY'S OVERALL POSITION

The agency concurs with the five recommendations.

AGENCY'S RESPONSE TO REPORT RECOMMENDATION

No.	Recommendation	High-Level Intended Corrective Action(s)	Estimated Completion by Quarter and FY
1	Office of the Chief Financial Officer should correct the Pesticide Registration Fund ("PRIA") financial statements to reflect the proper expenses paid by other appropriations.	Concur. OCFO corrected the financial statements to reflect the proper expenses paid by other appropriations.	September 24, 2013 (COMPLETED)

2	OCFO should ask the Office of Chemical Safety and Pollution Prevention to carefully review and comment on the draft and final versions of the PRIA Fund financial statements prior to their submission to the Office of Inspector General.	Concur. OCFO will request OCSPP to carefully review and comment on the draft and final versions of the PRIA financial statements prior to their submission to OIG.	March 28, 2014 (annually)
3	OCSPP, in consultation with the OCFO and other subject matter experts, develop a process that will provide accurate and timely allocation of Environmental Programs and Management expenses from other appropriations that support the PRIA fund.	Concur in concept. OCSPP, in consultation with the OCFO and other subject matter experts, will develop a process to ensure accurate allocations of expenses from other appropriations that support the PRIA fund.	December 31, 2014
4	OCFO should correct the PRIA financial statements to reflect the proper payroll liability amounts.	Concur. OCFO corrected the PRIA financial statements to reflect the proper payroll liability amounts.	February 26, 2013 (COMPLETED)
5	OCFO should closely monitor the payroll liability amounts for PRIA at year-end.	Concur.	September 30, 2014

CONTACT INFORMATION

If you have any questions regarding this response, please contact Christopher Osborne of the Office of Financial Management on (202) 564-5070.

cc: David Bloom
   Joshua Baylson
   Steven Bradbury
   Marty Monell
   Stefan Silzer
   Jeanne Conklin
   Richard Eyermann
   Paul Curtis
   Chris Osborne
   Sherri Anthony
   Raffael Stein
   Melvin Visnick
   Peter Caulkins

Maria Sorrell
Michael Hardy
Vickie Richardson
John Street
Margaret Hiatt
Robert L. Smith
Patrice Kortuem
Art Budelier
Sheila May
Janet Weiner
Janice Kern
Meshell Jones-Peeler
Dale Miller
Sandy Dickens
Sheldonna Proctor
Lorna Washington

Distribution

Office of the Administrator
Chief Financial Officer
Assistant Administrator for Chemical Safety and Pollution Prevention
Assistant Administrator for Administration and Resources Management
Deputy Chief Financial Officer
Agency Follow-Up Coordinator
General Counsel
Associate Administrator for Congressional and Intragovernmental Relations
Associate Administrator for External Affairs and Environmental Information
Director, Office of Pesticide Programs, Office of Chemical Safety and Pollution Prevention
Deputy Director, Office of Pesticide Programs, Office of Chemical Safety and Pollution Prevention
Senior Advisor, PRIA Implementation, Office of Pesticide Programs, Office of Chemical Safety
    and Pollution Prevention
Director, Biopesticides and Pollution Prevention Division, Office of Pesticide Programs,
    Office of Chemical Safety and Pollution Prevention
Director, Pesticide Re-Evaluation Division, Office of Pesticide Programs, Office of Chemical
    Safety and Pollution Prevention
Director, Registration Division, Office of Pesticide Programs, Office of Chemical Safety and
    Pollution Prevention
Director, Antimicrobials Division, Office of Pesticide Programs, Office of Chemical Safety and
    Pollution Prevention
Director, Information Technology and Resources Management Division, Office of Pesticide
    Programs, Office of Chemical Safety and Pollution Prevention
Director, Office of Human Resources, Office of Administration and Resources Management
Director, Office of Financial Management, Office of the Chief Financial Officer
Director, Office of Financial Services, Office of the Chief Financial Officer
Director, Reporting and Analysis Staff, Office of the Chief Financial Officer
Director, Financial Policy and Planning Staff, Office of the Chief Financial Officer
Director, Research Triangle Park Finance Center, Office of the Chief Financial Officer
Director, Cincinnati Finance Center, Office of the Chief Financial Officer
Director, Las Vegas Finance Center, Office of the Chief Financial Officer
Director, Payroll Management and Outreach Staff, Office of Financial Services, Office of the
    Chief Financial Officer
Staff Director, Accountability and Control Staff, Office of Financial Services, Office of the Chief
    Financial Officer
Audit Follow-Up Coordinator, Office of the Chief Financial Officer
Audit Follow-Up Coordinator, Office of Chemical Safety and Pollution Prevention
Audit Follow-Up Coordinator, Office of Administration and Resources Management
PRIA Audit Coordinator, Office of Pesticide Programs, Office of Chemical Safety and
    Pollution Prevention

www.ingramcontent.com/pod-product-compliance
Lightning Source LLC
Chambersburg PA
CBHW081236170526
45165CB00009B/3066